I0008189

PASSWORDS MADE EASY+

PASSWORDS MADE EASY+

PETER J. AMENDOLA

COPYRIGHT © 2015 BY PETER J. AMENDOLA.

LIBRARY OF CONGRESS CONTROL NUMBER: 2015903527

ISBN: HARDCOVER   978-1-5035-4995-1

SOFTCOVER   978-1-5035-4997-5

EBOOK   978-1-5035-4996-8

All rights reserved. No part of this book may be reproduced or transmitted in any form or by any means, electronic or mechanical, including photocopying, recording, or by any information storage and retrieval system, without permission in writing from the copyright owner.

Any people depicted in stock imagery provided by Thinkstock are models, and such images are being used for illustrative purposes only.
Certain stock imagery © Thinkstock.

Print information available on the last page.

Rev. date: 03/10/2015

To order additional copies of this book, contact:
Xlibris
1-888-795-4274
www.Xlibris.com
Orders@Xlibris.com
707946

CONTENTS

Thanks go to Chas Weiss and Eric Beck for assisting Comm 1 Solutions in the marketplace.

A special thank you to Doug Loucas, CEO, Comm1 Solutions LLC, for giving us the opportunity to experience the most interesting consulting projects we could have secured, affecting billion dollar assets.

Dedicated to the greatest technical mind of the 20th Century,

My Father

"The mind processes and remembers theory.
Once you have learned it, it is better than factual memory."

"I am extremely satisfied and grateful to have had my father, Dionisio J. Amendola, and Martin H. Tillinger, PHD in Physics, as teachers of science and Computer systems."

In memory of Business Partner Paul DeRico, Martin H. Tillinger Phd.,
And my "Dad" Dionisio J Amendola

Foreword

Peter started a long career in computer programming and statistical analysis in 1968 when his joint work with a schoolmate earned them a prize in the New York State ACM (Association of Computer Machinery) contest. They developed an artificial intelligence poker system that emulated the concept of the poker face. We all like to play card games, but Peter and Steven took game programming to another level at the tender age of 17, in a time when computers were at an early stage of intelligence. Peter and Steven were ahead of the curve. The routines developed, along with other contestant work, became in small part the genesis of computer analysis engines utilized throughout the world today. This work sparked a love affair with computers for Peter. Some of what was created was passed down from others that taught him the way to solve what was not initially well defined or yet visualized.

In recent years Peter would work in risk management areas related to computer systems and manufacturing, developing operational analysis methodologies with Comm1 Solutions, LLC.

Douglas K. Loucas

Introduction

READ ME FIRST!

Ok, you need to create a password, the dreaded word or set of crazy characters you can't remember a week later! You have too many passwords that you need to keep track of. Correct? So you make a written list of these passwords. This list then becomes something you also need to protect because it represents everything you are, own, and do. Why can't you have just one password? There are several reasons why we cannot have just one, but that is not the focus here.

It is a mistake to write down your passwords, so what then do you do?

When someone tells you to write down a password and put it in a safe place, it is a poor recommendation, because you are taking risks by doing so. You can't remember where you put your list for safe keeping, or it is not where you last put it. Because it physically exists, it is not safe, simply because someone may come across it. Lock it up and you have another layer of memory required to get to them and another logistic to be concerned about in several security analysis areas.

Does all this sound too familiar? Alright, calm down! You don't have to perform this hide and go seek exercise any longer. There is a better way. Follow my recommendation and you can keep the necessary information on you desk or in an open desk draw, unlocked, with an arrow pointing to your list that says:

"PASSWORDS - PLEASE TAKE THEM AND TRY TO HACK MY APPLICATION SYSTEMS."

The physical world, outside of your brain, is a dangerous place. So why use it for security? This is why the greatest spy systems in history relied mostly on the human mind to store "intellectual property".

If you have not yet found your list, I have a brief message for you. Sit down and read my book. In "Passwords Made Easy+" we are going to give you a system for creating codes that will tell you what your passwords are for every sign on application you use. And you will not have to understand but a few things to determine your passwords, no matter how many passwords you have. You will translate codes or hints to your passwords with programming and association theory. All you have to do is understand the theory of how to do this, not remember what your passwords are. After a while your system will become ingrained theory that will not fade from your mind until maybe you are in your 80's. Notice I said mind, not memory.

Yes, you can determine your passwords by using an association formula, based on the brain's ability to associate, not using your memory. I am not talking about those pesky challenge questions when you loose your password. Those should be easy to remember, such as your first car. Mine was a Dodge Dart. The problem is that these popular questions are not at good security levels, with only a simple, direct, first level connection to facts. Too many people know this information about you! The more popular, or worse, famous you are, the more people know this information.

In some cases you can't retrieve your id or password because you can't remember the answers to the challenge question, or the application requires some information you can't find, such as the account number, in order to reset a password. By the way, the Dodge car ownership is public record. If the police can see even this old information, someone else that wants it bad enough can retrieve it.

What you need is a great password. Why just randomly choose one out of the blue sky, so you can then forget it a week later?

First rule: If you told information you want to use as a password to anyone, or anyone knows this information about you, and many probably do, you should not use it for a challenge question, and especially not for a password. I was recently asked on the phone for my wife's challenge answer, to get product service. I was able to answer it exactly, the first time, without hesitation. The question had nothing to do with her family or herself personally. It was just something or someone she would have admired. I know her and what she would choose. The best associations are things that happened to you as opposed to things about you or your favorite people or things directly. If you can make an indirect association about a person, thing, desire, a concept, history, science, or language, that will work better.

My system will insure you don't see those lost password challenge questions because you will always determine your password first. The scripted list you create on paper for your passwords will be coded hints that only you can figure out. You can make 10 copies of your list and give them to anyone without worry that they will ever be able to make use of the list, let alone make sense of it. I have my coded list displayed on my computer favorites for all to see, fully knowing that the information is absolutely useless to all who look.

I think at this point there is some value in saying a few words on the subject of challenge questions. I do believe in them because there should always be a backup. A backup concept should be no different for security protocols. Please consider questions that few people know about you. When you form the answers to these questions you should spell the answers phonetically, not correctly, or place imbedded characters, especially if the application has a strong security requirement. This however is not an absolute requirement. Example: use a "k" for "c" and visa versa or use a $ sign within the word. Always try to understand that the functional purpose of what we are learning here is not to have perfect security, which I believe impossible, but to figure out your passwords when they are needed. Always use at least suffixes or prefixes (suffixes or prefixes can be numeric, special characters or letters).

What I would like is a way to dynamically create challenge questions and answers using my own subject matter. That is basically what I am proposing for password creation.

Before we continue on our journey, I should define a term. There will be a breakdown of information associated with a password. This is done by the use of elements. An element is a segment of a set of password hints that tell you what your password is when they are put together. This means that more than one hint (element) can be included in a password description. The element is the smallest part of a system's data that will allow you to translate back to your password. You determine the length of a hint and the associated password element, and the theoretical complexity. It will be part of your system platform. You will write the element associations down in a scripted language (your list is comprised of these sets of elements). I am not talking about variations of a fixed theme. That is dangerous and easy to figure out. Elements are "mentally" descriptive and not static values or variations, nor are they direct factual associations.

If you wish to make your passwords complex, this system will work for many password element sets also. More password sets in a password sign-on will however have diminishing returns and potentially negative effects in figuring them out. I never go beyond 2 password element sets to make up a password. What I am saying is that there would be as many as two passwords strung together to become one password for a sign-on. Each hint set in a password description can have multiple elements associated with it.

The strength of your password will depend on the strength of your system's choice of password subject matter, its unique underlying theory, and your use of mind clicking hints that you and you alone understand and can easily and definitely equate to an answer or "password". The theory could be the subject matter's scientific theory, and not the theory of your association methodology. The elemental hints could be 2 to 3 levels removed from the password answers (indirect or connected theory). You will develop a language that creates descriptive functions as well as subject matter hints or translations.

This sounds like a difficult thing to do but once you start thinking and developing your own system you will look back on this day and say, wow, what was I doing all those years, randomly choosing words or sets of prefix/suffix and embedded numbers that I could not remember a week later.

Let's go back a bit. When you started writing your passwords down literally, you exposed all your security. Never write down your passwords except when you first create one. After you have encoded it into hints, cut it up, and eat it! I am just kidding. Simply throw the shredded paper away.

Getting back on track, you might think my method will not cover the latest standards for complex passwords, but it does. The length you want a password to be must however meet an application's minimum and maximum character requirements. We will look at this in the standards chapter. An application minimum of 8 characters is usually the requirement. Some are shorter and are not good at that length. Be concerned with applications that use a maximum of less than 8 characters.

I will also show you how to update these passwords so they do not conflict with prior passwords used in rotation over a year using 12 passwords. Some applications require lots of uniqueness when compared to prior passwords. How about 12 different passwords for the same subject matter!

If you run into an application where my system can't work immediately, simply make a documented script exception using the language you develop. Again, what I have found is that the minimum or maximum characters allowed, in some applications is too short for my needs. That is where I have placed exceptions in a hint. The hint exception is also not literally specified. If you do run into a short password restriction of under 8 characters, I suggest it is a poor security module in that particular application. The short password restriction is really not a problem though, as our primary objective is to figure out a password, and not change the world of passwords, just yet!

OK, you will learn that it is the elemental hints that will tell you what your password is exactly, without the password being revealed to others, spoken, or written down either. And the elemental hints will always equate to one set of contiguous characters (the password). The hints can be 1, 2, 3 or more elements, or multiple sets of elemental hints for one password. You will see that the word "elements" is loosely used for a single element set or multiple sets of elements, forming a password or multiple passwords within a particular sign-on password. Your choice.

You can write your hints down for a particular sign on and no one but you will ever be able to figure out your passwords, not even me. Wow, a sheet of garbage for others that means the world to you.

Yes, you simply encode your password using your minds ability. I will explain.

By now you are probably getting annoyed: "what the heck is he talking about? It can't be that easy, and I am not a mental genius". Is there a middle ground? Yes. It is precisely what I am doing by telling you things more than once. This is what you need to ask yourself a few times as one of the tests of your system. Is it catchy and will is automatically come out of your mind. It is not something you have to memorize to translate. Your brain's mind (your processor) will do all the work thru conditioned response.

Think about the wonderful logos of corporations that, without words, convey an entire concept, and even the actual name of the company. This is a subliminal association. We are going to do this to some small or great degree depending on the requirement. It all depends on your ability and needs.

For those of you that want to go to Chapter 1, I will see you on the other side of knowledge. For those of you that like to read and think, I applaud you for taking the time not to read through! This is like playing through in golf. The golf course holes are like the chapters in a story. So if you want to associate golf with reading in your passwords I just did it for you, now think about the two. But you really should do this with life events that happened to you.

In closing, the key is getting your system correctly designed for your needs. That has to do with the proper subject matter and your skill in programming or "scripting" it. Did your brain just tell you that you heard or read that before? I did however use different associated words.

I was privileged to work in military contract with a secret security clearance. Only I know where I worked on special assignment, and I never told anyone a single address in a city because it was sensitive information. These particular places all change location anyway. Ok, that information is not part of my passwords. Why, because I was not the only person to know this information. I recently ran into someone that worked in the same area of endeavor as I did and he knew about these locations also.

You may have used some of the techniques in my book but I ask that you make a system out of it and develop a more robust platform for your passwords. So please stop writing down your passwords. Just design and then write down the theory using a scripting language you will develop from Passwords Made Easy Plus. Great stuff, writing your own translation language!

Have you found your list yet? If so, we may be able to reuse all or some of your passwords if they have a great association with your personal experiences. You will however need to write translation scripts for them.

OK, you know that security and your poor memory are the problems, and not the actual password itself. Your passwords may be good ones, but just the wrong ones for you to remember. Passwords Made Easy Plus will take memory out of the equation. I think you will have fun with this!

Chapter 1

THE HISTORY OF PASSWORDS

A long time ago, in ancient times people used pass-codes to allow people to pass through their domains. Maybe those travelers carried some sort of object to tell the gatekeepers who they were. These travelers sometimes fought off those that tried to steal these objects from them. These objects were considered valuable keys to travel and prestige. Sometimes the gatekeepers would fear the traveler because the object was a sign of the traveler's high position, or some supposed super-natural ability.

Does some of this sound a bit like 21st century terminology? Well, we learn from history, as what we do is passed on to us, unless it is newly created. The question is always if we are ready to accept and make proper use of life, agreements (standards), and new technology?

We may not have made proper use of our ability to create good pass-codes or passwords in the early days of computers. With all of our technology, we still did not do justice to the ancients (gatekeepers), those early historical users of the now famous words ("who goes there?"). Maybe it is time to improve the science of passwords.

How about electronic ID's? If everyone could afford an electronic id maybe there would be no need for passwords? I think not, as even electronic secure id's can be stolen temporarily, without anyone's knowledge, used, and then returned. Physical security can be breached easily where the theft target is mobile, traveling into unprotected areas. For that reason, a good password is still needed when using an electronically calculated secure key methodology. Imagine being at a critical point in a project and loosing your electronic id. Everything has drawbacks, so exercise good practice, and have a backup strategy. My practice will be the best you will find. As you read you will create your own system containing unique subject matters for your passwords.

I just decided that electronic smart id's are risky. Why? Because they are physical security that can easily be lost or stolen. Would you not want a security that does not exist except in your brain? Look at the cost and complexity of each security.

For most of us, the best security is kept in our brains. We just need to culturally learn to be good agents of security to ensure that we don't give up our information.

Here is some history that you may also have experienced. Many years ago I used primitive passwords and wrote them down when it became obvious to me that they could not all be the same password for every application due to restrictions and business requirements. I ended up with many different passwords that I could not remember. This became worse as I got older. Which password was for what application, I could not remember? At some point I had almost 30 passwords. Many of us have created over 100 passwords not including those temporary two week periods while testing new applications.

You should assign part of a system for those temporary passwords and also not use a critical password result for temporaries. Yes, you should build a separate strategy for the temporary passwords. I use half of one two-part system for temporaries. This will become clearer later on, where we associate password sets for work, play and personal, and use them together as double passwords to enrich the strength of the password for critical applications. Remember never to place the most critical personal passwords sets in a work or play type password. Yes personal passwords are the most critical to you, therefore don't use the same passwords for all application types.

To continue, I found that I had to create new passwords, many times, due to restrictions. They could not be too long, could not be too short, and could not repeat the same characters in succession. If we could only develop a standard for passwords! See you in my next book on that very subject. A subject that might very well become streamlined by the techniques I am proposing. I will touch briefly on standards for complex passwords, and what I think of them.

Let's come forward in history about 5,000 years. In recent years internet applications have implemented site-keys which I view as a great thing. Security experts believe that there needs to be a more comprehensive way of not only making sure the user is in the correct place (web site), but also a need for being able to specify a hint that is presented to the user with the site-key in the sign-on process. This is good, as the scripting language that I propose can be shown within the site-key screen as the user enters the password, or what some sites call the "pass-code". I will use the word password to also mean pass-code going forward. As more and more site-key methods are incorporated into applications we hope that the need to write down our scripts may vanish? No! But at least we can put our list away and not have to reference it, unless the application lost our site-key information or we are rebuilding an old application. Another reason could be a new version of the application's screens, where the site-key hints are not retained. In any case, please keep a list of your hint scripts as it is a good practice to have a backup of everything you do. Remember it is not the passwords that you write down! It is the descriptive hints as a language that you will write down. If your system is good, you will not have to wonder or worry ever again. You will need to document the parameters and format of your language.

Recommendation: If an application or web site uses a site-key, take advantage of it, from the beginning. Do not hesitate.

It's time to remind you that you are creating a system for you to figure out your passwords, which is the primary objective. Do not forget that you are also protecting yourself from people. Concerns about automated computer systems that will try to get around your passwords are of lesser concern for this discussion. My system will however also help you with automated pass-code ciphers. Remember not to spell passwords correctly, as they appear in the dictionary, if they are one word, especially if the application requires a very strong level of security. You can cryptically specify miss-spelling in the language you will develop. For personally sensitive passwords, this level of spelling protection may not be as important as most hacking is done against business databases, and not your user-id interaction. Database passwords is a major subject matter we will not cover here, but you, as an administrator can use this system to create a password for work database access.

Chapter 2

STANDARDS FOR PASSWORDS

By now we all know the drill for what is a complex password. We should enforce this as a basic standard. This would eliminate the short and the obscure password requirements, thus allowing the full use of a language for developing password hints, with as many elements as we desire, being concise with our methodology! I also recommend that your passwords be more that just the minimum complexity. Complexity is also in the subject matter relationship to the password. So don't go overboard in creating a long monster password. Done properly you can make a simple word as complex as the story of the lost ark.

Again, remember to be as concise as you can with your language and passwords. Abbreviate as much as possible, but abbreviate well. You can also create your own standard abbreviations for password hints. The relationship of the hint, back to the password, needs to be something that is ingrained in your mind.

OK! Notice that I put the words password and hint in a specific order. I then used the word "back" to mentally tell you in your mind the order of the words "hint back to password". The word translate was not used, but you understood from the relationship that translation was what I was talking about. This association is only a simple example. Since it is not an event in your past, or something important to you, it is "not" the type of association you should use for a password. Although it could work, it is not a completed association. It could only become a repetitive knowledge at some point, after 20 or so entries of the password. It needs to also be tied or associated to something memorable in your life to make it permanent in your mind.

I will continue to remind you that you will be developing your own system. You will have the knowledge of how to create a system of your own very shortly. Your system will be unique, and your language will be your own version, helping you to create a translation standard of your own. Your personal standard!

For now we will discuss the password side of the requirement. Remember that you will be developing passwords based on your subject matter, after you develop your systems, and the integrated language that goes with them. It will help you to create better passwords if you follow the standards path I will show you, not only for passwords, but for subject matter and language. I will make a list below. Your developed language should remain the same for all your systems. If you want, you can create more than one language or variations of one language, or function formats, but I do not recommend it. Remember, concise and consistent as a rule, will keep you out of trouble. I continue to caution you, as this will become fun. Don't let the fun get in the way of your primary objective, to figure out your passwords, not having to remember them ever again.

A list of some of the concepts you should research and come to understand, as you become the developer of your own standard for the creation of your passwords follows. Try to understand the meaning of each area.

This is the core of terms that you should master, and all are integrated in your theory:

1. System – The structure of the password hints to tie your subject matter together.
2. Subject matter – The best subject matter are subliminal and unobvious associations, and very personal to you
3. Language – How you write the hints for your passwords
4. Password – The resulting answers to your associations and neither are ever written or spoken
5. Standard root password – Can be the shorter, or longer, for a set of the variations, developed for your system, over time. You can reference it as the standard, with an abbreviation, that only you know as the standard. This abbreviation will

work because you also have the standard written in your list, documented as the default standard. The root is not necessarily the shortest variation in terms of the number of characters.

By no means is the list above complete, as there are probably well over 50 sub-concepts you will deal with when creating a system. You will deal with sub-concepts such as static and variable. Sounds like a math or science book. You will not only deal with the metrics within the language, you will also deal with an "any subject you wish" arsenal as part of your relational theory when putting together hints for passwords. I want you to think harder than you ever have, to create your standard for all the core elements above. Standards should be considered for all we do with this exercise, not just the password. It will be your standard to carry you forever. This is a programming language that will never be phased out because you will create and control it.

A standard is a commonly accepted or registered methodology for documentation or for performing a task.

The most important part of the list above is the unique subject matter and how it relates to the created password's real text values, which are the answers to the equations for the passwords, the special characters of the password, and the suffix or prefix, added to, or imbedded into the password. Associations are the underlying theory, never written, that is in your sub-conscious mind. Visualize the subject matter physically, as it happened and in characters, something we will touch on later.

Think now about a good system, so you don't have to rack your brain trying to remember a password a year later. You will have to write down your script codes (language) however. You do this after you have developed your language basics and documented the format of the language, just as I have done with the example in Appendix A. Go there now to review the example. You will modify and expand the language for a period of time initially.

Getting back to password standards and for lack of a better password standard, the most commonly held opinion is that a password or multiple passwords in a single password (what I propose) should contain at least the following:

One upper case letter, numbers, letters, and at least one special character (two special characters is ok in my book).

The upper case letter is not always necessary in my book. Notice that I did not say anything about repetitive characters not being allowed. The use of repetitive characters is ok in my book, but I would not use something like 1111111 as an entire password by itself.

A serious problem with passwords is the lack of an enforced standard, similar to a protocol standard. Companies and organizations create their own unenforced standard, thus affecting consistency, which is important to creating user-id and password standards on a global basis.

Here is something to consider. I have written several data analysis computer programs in my life. Always give a litmus test to your password to ensure that it is not, for the most part, in alpha, numeric, or bit binary numeric ascending or descending mathematical character sequence (1,2,3,4,5, etc. or a, b, c, d, etc.). Also please do not use a random number sequence generator to create a password or part of a password. That would defeat the purpose of a system, which I think is a better way of managing passwords, giving you systematic control of your password content, making it easy to figure out your passwords. Remember the primary objective or prime directive, ha ha!

I am not saying that an electronic smart-id type of numeric generator is bad, just unnecessary for my methodology. It is just bad for you to use as the sole means of accessing your applications. These electronic smart-id generators ensure that the results are random and have no pattern. There will also be a password associated with electronic id usage. We are working in an area of science related to passwords however. Getting the password correct for your purposes is the primary objective. You are not supposed to pick a word or set of characters out of the blue sky, or one that does not have much meaning or sufficient secrecy. This is a mistake we have all made in the past. Using a generated password is equivalent to out of the blue sky. So if you use an electronic id, a good password system is also necessary.

Consider that there are programs that cipher passwords in an attempt to break password security. Keeping your passwords away from only words and on a track that does not use correct spellings is a plus. Again however, it is not the primary objective, which is to get your password correct when logging-in to your applications. The spelling can be correct if the security requirement is not so critical. Hint - the use of double words as a password helps in this area.

Your level of security requirement will secondarily dictate how complex and masterful your subject matter relationships need to be. You need to balance all this against the average hacker, not the United States Pentagon, FBI, or the CIA!

Chapter 3

Choosing a Platform for Your Password System

(Structure)

Before we begin, it is important to understand that you can make your system as complex or simple as you wish or need. In either case you will have a super password that will be secure, and one that you will be able to figure out. Structure is how you physically lay out your system.

A platform is composed of elements that are put together in an order or stand by themselves, each as a single element / password relationship (the simple case). The platform should have good structure that can be inverted, numbered and then placed in another order by specifying the order in the hints (this can also be done for a single element/password set of relationships). The platform is what you derive your language script from. The language will allow you to write the element order, or change it, so don't be concerned. The system has to do primarily with the subject matter and how you choose to use it (integration). Always use two password sets for your system. You can use one in a simple password, but the use of two related or unrelated subject matter sets will give you flexibility.

Here is a simple example:

The script might say thirdpart, firstpart, secondpart or thirdpart, firstpart, skipping the use of the second element set. I am using commas and adding real words to the description of the script, but your script should represent continuous resulting password characters, and be as concise as possible, to fit within application or web site-key character limitations. Remove the commas from the resultant password. Your language can use the commas however. I will give examples later. You need to think about the password relationship and also how you are going to write or "develop" your script for it. I have provided a basic overview of a script language in appendix A.

You will develop your own language, which is what I want you to do, to make your system unique and impossible for others to figure out. You will find in your mind the ultimate language to fit your systems, just as you will develop great subject-matter associations.

In a simple system I had once upon a time, I used 2 to 3 elements that I used singly, doubly or in triple. These elements were defined in my hint system as first part, second part, and third part. When I write down the single hint set script I write firstpartxx or xxfirstpart or secondpartxx or reversefirstpartxx (the last hint tells me that the password is in reverse character order except for the suffix, while the first two hints are simply taken in normal character order). If two passwords together is your need, then firstpartxxsecondpartxx may be the script for the password.

Ok, what are these passwords? I know them but you will not figure them out even if I told you they were the first and second parts of the system, reversed for the last password.

You can design a language that states reverse-all, in some way, as the function and have an as-is function for the suffix. This is only the structure as the key is in the developed relationships that are not written in the scripts or anywhere else. The relationships remain in your brain. Note that you can develop only one positional (first, second, third part) system unless you number them 1 and 2 and so on. Your subject matters could also be theory or concept associations and not positional.

The first rule, and the only rule, is that if you ever uttered the resulting password elements or numbers, don't use them for this purpose.

Note that the parts that form the positions of the above system are a simple relationship to subject matter that actually had a positional relationship originally. The relationship is in my mind, and it comes out using a subliminal release. Our mind can construct a visual image of information if we learn it and then picture it. Do not memorize it. Yes you can image the characters in your brain. You know the adage of a picture being worth a thousand words. Well why not cultivate the science and use it? This is not memory. So this is one technique you should incorporate into your methodology.

The mind's release comes into play when we view the information with an associated visual image to help trigger a conditioned response. You can do this in many ways. Learn to see the letters in your mind as a minimum technique. You simply need to see on a virtual screen in you mind the information as a picture of letters, and then you will have a hard time ever getting it out of your mind. This type of visualization is only one technique for triggering results from associations. It is a good way to do positional associations because they don't always have a real relationship to events, just theoretical order (first, second). When you visualize both parts together a few times, the first and second parts also become associations to each other in your mind.

Structure is the word I use to describe all techniques because we can see structure. What about organizational structure in a corporation. If we train ourselves to visualize theoretical elements such as an organization chart we can trigger associated concepts related to the blocks, such as "upper level management", as a result. Not all associations require visual aid, but it only makes it better if you can put an aid into your mind as it automatically links it with the information on the way out. We do this without trying when we think of the character of a person which is an association. We see the person, the face, the body, the expressions, as it is automatically triggered. We just need to learn to intensify this ability and also make it easier by choosing the strongest associations we can find from our experiences. My early life reading was rich with history about cultures, their customs, costumes, lifestyle, and more, creating images of centuries of events all be it in still pictures.

Ok, let me explain further the first part, second part relationship and the associated password(s). I have an old password that was somewhat complex from over 25 years ago. It had two parts (in an original positional order). I still use this firstpart, secondpart, subject matter as part of my first system.

It is now a good time to tell you that the original password from 25 years ago had a format (structure) that was randomly given to me by an application. I have not used it for 15 years, but because I developed a system from it I know the format and the special character that was randomly placed in the middle of it by the application generator. The way I remember it is scripted as firstpartconsecondpart. This means nothing to you but I used it so many times and then built my first, and only, positional system from it, that it can't leave my brain. No one other than me ever knew this password. The "con" in the middle is the relational hint for the special character. It is also a function that says connect it to the secondpart. It serves two purposes. This is the consistent type of thinking you must develop in writing a higher level language.

Ok, I am going to give you my actual relationship that I will now never use again. The relationship of "con" to the element is "concatenated". When we concatenate text elements together in computer programming the end result is the "addition" of two text parts together, side by side, no blanks, no character inserted between the parts. I am changing the character insertion rule however. The special character resulting password sub-element is a "+" character because addition in math is specified as a plus sign. Do you see how removed the two concepts are, the "con" and the "+", yet related? The result is password1+password2 with the plus sign as part of the resulting password, or you can remove the character. I don't use it any longer (the plus sign), so that is why I revealed this to you. Yes, "con" is a function that prefixes the secondpart hint element. I would bet you took the "con" to mean "pro versus con", con being the negative or opposing. You will never know what first and second part are translated to however.

I sometimes mix two systems and put the elements or parts in any order I choose, when creating or changing a password. I also use the system to modify an existing password thus creating a variation in both the password and the script (not the language). Just be careful for your mental capacity's sake. I knew the old firstpart+seconpart password very well, so first and second part was very easy for me to release from my brain. It is ingrained in my mind just like Columbus discovered America in 1492. The actual passwords will forever remain unwritten.

I caution you not to rely on your memory in creation of relationships. This firstpartsecondpart password works only because I used it thousands of times in its original conditioned form. You still need to use a well know (to yourself) knowledge base for relating the subject if you choose this positional technique. This is different from using position as a function! What I am saying is that the first and second parts must be ingrained in your mind for many years, especially if they were obtained at different times (mine were together in time). You should consider if this simple order relationship is good for you.

I now have better methods. Remember memory is bad, relationships to people are good, and relationship to real personal events is the best. In effect this old system above also had an event because I created one in my mind. I can still physically see myself at my computer, in my special office, typing it in. I see the lighting, the custom built-in desk in an alcove, constructed into a 3 foot deep closet, by 5 feet wide. I can also see that computer and every time I think of or see the password hints, and the room, the password rolls off my fingers onto the keyboard.

Remember, if you have the written hint you know the order (by subject matter) and if it has a special suffix to vary the password from month to month if required, you have that all in the script. This suffix change method is only if the application allows a simple suffix change from previous passwords used. Better to use varying special characters if allowed by the application.

If the application does not allow the suffix change as a valid change, place the real password characters in reverse order with a different suffix as a suffix or a prefix, as needed. You can reverse the prefix, or if the application does not allow that, jumble the word part with a numeric formula, each month. You can also split the suffix or prefix and imbed it into one, or both of two element sets using a position function. These are just a few ideas. There are many more. Just work with the application rules for password security.

Example:

"Hintxxhintxx split suffix george burns" translates to Passwordxpasswordx, or literally James9julie9 (the real password), where the x's are 2 numeric digits split to one each by the split. George Burns lived to be 99, although everyone thinks he made it to 100. Remember that the capitalized P means to capitalize the password's first letter. If all caps, then all caps for the password (letter part). Only you know that the xx stands for one character each and the character suffix could be a letter (your language), a number or even a variable special character over the months or quarters of the year. The password could be one or two passwords depending on the hint or hints you choose to tell yourself as I have just done. The finished product hint for the example above would actually be Matexxchrushxx split suffix george burns (no blanks), again literally James9julie9 is the password. If you have to rotate passwords on a monthly or quarterly basis then the 99 would not be practical, but would work for a static password requirement.

Ok, the relationship: James was one of my kindergarten friend/ mates long ago. I saw him grown many years later, and I remember this sighting as part of the association, as I could not call out to him to say hello. Julie was my high school crush that got away from me. I still think of her with passion today. Start thinking about the past, and the future. Present or current events are poor choices for subject matter, as it is what most people are thinking about now.

It may sound complex, but once we go through a few more examples it will start you thinking, and you will see how great this is. And it will serve you for the rest of your life. It just has to be well known to you and make incredible sense to you alone. People often joke about being "great in their own mind". Well in this case it would be a blessing.

The choice of the passwords is of course the most critical part of this system in addition to the relationship (association) which is not ever written down. Only the hint/equator is written down. The relationship remains private to you. You need to think this way to create a truly, closed to any guess, type of password system. But it must be placed into a structured platform similar to org chart rules as one example.

Your platform can choose to reference a (non positional) person or persons and the password would be something unique to that person, so your elements can be George, Frank, Paul, but these hints mean nothing if written down. Don't use public figures as hint relationships, as someone can then guess Bush or Washington as a real password for George.

A better example would be Steve, your cousin that had only one arm, and in your mind alone you considered him to be a hero. The hint "steve" would then translate to the password "hero". And if you wrote the hint as Stevexx, the password would be capitalized Hero with a suffix of say 01, that you could vary over the months as needed, making the password for January, Hero01. Your hint would not state 01 but xx or ss or pp or xxx or sss or ppp. A good way of being cool is to use one x and it means 3 characters as a suffix or set of imbedded characters. Only you know that and you state it in the script, or not, as you need to in some way ("suffix not literal", three words for three characters). So the suffix x is three characters, not one. The variable suffix is a simple example that would have to be more complex for 12 passwords in a year.

It is what Steve may have done (an event) that is the association in this case above, that you need to have visualized. That then makes the connection of Steve being a hero and keeps it in your mind. This image needs to be a strong one in your mind. Better if you actually witnessed Steve's heroic act (the event). The password could also be something about the event and not "hero".

These examples are singular and somewhat simple. You need to become as creative as you can. You will have to document your language and the functions you will create. Yes you will also create your own functional operators that act upon the subject matter.

Ok, the key is not to be literal with the hint, but create a relationship that only your mind knows about. Two conceptual steps removed from the password answer would be a great objective.

To give you another idea, suppose I wrote a script that had steve+steve in it. It could mean hero, but in this same system's language, it could mean a special character ":" which is visually two periods. Steve would be a special person. Steve would remain as Hero. evetS would translate to the four letters "oreH". Notice I said no where in the script reverse. I used the theory in place of the functional word reverse by spelling the hint in reverse order with an upper case letter also. You can also develop a jumble of the hint that would tell you the position of the password characters. For that however your password (alpha part) and hint should be the same number of characters.

What about a suffix. It can be as many letter or number characters as you wish that you agree upon as part of your language. So xxx can mean 1 or 11 or 111. Heck, decide on 01 or 001.

What about a special character requirement? No problem, choose a special character and then simply place or develop a method for noting it. Don't use blanks for the special character. I will not reveal how I do it as it is part of my system and I don't reveal certain methodologies. Sorry.

Here is a simple example and way of doing special characters. Choose a special character, and anytime you need one, specify the name of your past love. The relationship would be that they were "special" to you. If the person's name was not special to you it would mean something other than a special character and you would use it for a different relationship. Just don't reveal the method. The person could stand for hero and not a special character, only you know this however. A special character could be a special object to you also. So a favorite "guitar" could be a semi-colon (;). Can you just see that electric guitar make the squeaking sound in the tail of the semicolon, and the image of an F Clef which looks like a semi-colon. It has to be your mind's relationship, not a defined relationship by me, removed from the hint's exact dictionary meaning, and not directly attributed to the password. What we look for is a catchy image or phrase. These are just two methods. Think hard, and rest easy next week when you need your password.

In closing this chapter it is very important that you create a default or root-standard password for every system, which will allow you to use an abbreviation and speak it in a crowd to your wife, if you love and trust her! It will then also allow you to build variations from that root or the associated abbreviation. This is not the only reason for an abbreviation.

Start thinking out of the box as variations are important to keeping one step ahead of someone that might be looking to snag your password(s). You can also use the abbreviation and add a variation script to the abbreviation that is from another password system all together.

Did you ever think that just the words complex-complex could mean a special character ":". In fact, what would be better is "complex-half" translating to a semi-colon. Written languages are vast. Just use words other than those I have just used. There may be hundreds of words that can be used to mean the character ":". How about double up for colon? And two periods could have the hint "suffix double down" "..".

Here is another idea - colon_colon is really two periods, a colon laid down on it's side, which looks like two periods. The underscore tells you that the special character, that is in your mind, is at the bottom of the text line, laid down. The translation comes out to be two characters ".." and not a standing colon. All part of your language standard.

If you can't get your first system to accommodate an application's security module, use your second or third system with another knowledge base (subject matter)!

Don't forget to start the suffix with the month or quarter you are in, in order to satisfy the application's periodic password change requirement. Some systems require 3, or 4 changes per year. Some use 12 per year. Create 12 passwords from the same subject matter (discussed later).

Remember not to use the exact examples in this book! Sorry. Besides, if you use Julie, you will forget her name as she means nothing to you. To me she was everything, at the time, and someone I will never forget. You have passion for a few people in your life, although you will love many people in many ways or degrees. It might be a loving admiration, as an example. There is an association that has three parts. The hint might be loving, the association James, and the password admiration. James would have to be a special person in your life. You would have to carry, in your mind and heart, this admiration, to make it work for you. In my case no one in my adult life ever knew James, not even my parents. This is a concept with an event in time where I could not say hello. My mind still suffers today that I could not call out to him. Just to say I remember you and I wish you all the best in life.

What this all comes down to is structure, be it visual, physical, or in theory. It should have structure, even if it is simply virtual structure. Structure should innervate all the parts of your system which then needs a language.

What if all I have laid down for you is not enough to get you past the application's security requirements? It is so strict that you can't get by the periodic change of password each month. The answer is to simply create a table or array of password hints, usually for one subject matter, that has 12 different associations. Each month there will quickly be a recycled hint waiting for you. The following year you can vary the passwords based on variations if the application requires it in some way. It is best to use the same system and subject matter for this application's requirement each month. Commonality in subject matter helps to create an associative tool.

Chapter 4

PASSWORD USE

Now that you have an idea of how to create a system, you should consider that the parts of the system should be used in different categories. For example I used first part for work and second part for personal use, making sure I never cross the two.

You can take two systems and have one borrow from the other, thus putting them together as a single password set. You create a dual password system (two in one, as was first and second parts for me). This is all your choice. You should try to keep a system's subject matter the same or similar to prevent like-lettered words from other systems affecting your mind's associations. This could be a problem if you create many systems with many subject matters and associations. So be careful not to defeat the purpose.

If you have to give out the work password for an emergency, which is still a no-no, and we all know it happens, I would change the first part meaning shortly afterward, using a different subject matter relationship set or equation, keeping the same subject matter. Don't simply reverse the order of the text, but do more with the prefix and suffix or another association from the subject matter set.

This need to change may not always be necessary. Changing 20 or 30 sign-on passwords might also be time consuming. Only give out a password if it is not that sensitive, or you trust that person and they have a need to know. As an example many of my wife's children know my email password. Remember this is primarily a system for you to figure out your passwords, that you have written down the hints for. If you have 30 places to change I would not give the password out. Remember it is the password, not the hint you are giving out. If a person that used your password subsequently sees the hint, they should still not be able to figure out the password. It should (the hint) mean nothing to them. If you are concerned, take the suffix and split it so they could not use the changed password, then change the hint to add split to the hint language script for only the password/ application in question. In this way you can use the old password for other applications that this person has no knowledge of. The best thing is not to give out a password. If I think I have a problem with a loose password I change all my passwords to another backup system (subject matter relationship). I have a couple of systems that I now use.

You can use the same password in a few like applications, such as banking for example. This keeps the methodology tied to specific systems and applications for your control and you tend to associate them with the password by type of application. It is automatic, as you see the bank buildings as part of the association. Try it sometime by placing the image in your mind. When you think about the bank, does the image of the local bank come into your mind? Some people cannot do this because they are not wired that way!

Ok, except for what I just said, I lied. There is a second rule. Vary the hint / password result for a system across your Id's. Remember that you can put slight variation in the banking passwords, and your script will tell you so. If your password is excellent then you can use the same one on all your banking. It is your choice. I use two variations in banking passwords. That is just me. Don't use the same password over and over again for everything however. This protects other applications from those that you gave out a password for.

Here is an idea. The use of a double password will challenge the memory of other people, who somehow obtained a finished real password. I have some 21 character finished product passwords (consisting of two complex passwords). Remember that you give out a password and not the script that is used by you to translate. In this way they can forget it and have no hint.

Woops, another rule, don't give out your passwords. In a critical application you never give out a password. Use a temporary.

A good side effect of variations is that you will only have to change one or a few passwords if they need to change due to compromise or suspected compromise.

Alright, another rule: If you suspect someone might have your password, change it. Risk management principle states that you perform a mitigating action based on logical necessity and not emotional fear. If there is any reasonable potential based on your investigation, just do it! What? It means that if it can happen, there is a chance it will. Password protection should be a discipline based on engineering practice.

To what length do I go to in changing a password? Simple! Keep the system and just change the relationship for the subject mater, thus changing the password and script. You need not go to a drastically different password system or structure to make a change. The change needs to be more than a single digit.

Chapter 5

CREATING A SYSTEM

Now it is time to create your system. It is easier to remember how the system platform theory works than remembering the actual passwords.

The mind processes and remembers theory in a special way, through associations. Once you have learned it, it is better than factual memory. You don't even know you are applying it in solving a problem when using it. For this reason I suggest that you use scientific theory within subject matter, in some way, in your system. This does not mean it has to be the subject matter. You do this by employing a scientific principle in the translation of the password, not necessarily as the subject matter. If the subject matter is scientific, that is acceptable, only if you follow the rule of keeping the meaning or association in your mind. I have tried to demonstrate a bit of that in our discussion and some simple examples. Why does it work? Because it is conditioned response and when you learn something in conjunction with other things these things are linked together. When one is presented to you, the other is automatically presented to your conscious mind almost as a look ahead record, but with a theoretical association that is never spoken, tying the parts of the concept together. This is the basis of Passwords Made Easy +.

I think it is time for me to create a new system myself. Not, as the children say today. My systems are good so I think I will just keep them. I could create a much more complex system than I could 15 years ago, but why not enhance a good thing and let it continue to grow. When I need to create one I will. Remember that need is what should drive your development of systems.

Did you catch the simple reverse relationship in the use of the word "not"? Such a simple word that means so much when used in the way it was, countering all of what was said in jest beforehand.

You should now start taking notes for yourself. You can place as many subject matter relationships into your system as you wish. Try to get as many as you can on the same subject matter so you can easily classify them should you come to have many subject matters in a few systems. It is your choice, as always, as to what element is part of what subject matter. So subject matter is actually at the core of your system and platform.

You may need to create three to four systems to cover all the various needs you might encounter. I am hoping that you are smarter than I am and will accomplish it in two to three systems. Maybe you will only need one system. I can't know that for you. You may not encounter as many situations as I have over the years.

Let's get started. Before we begin, we should understand that in one system you can create as many as 12 relationships or more in one or more subject matter areas that are related in some way. This is for those applications that require strict password change on a monthly basis. The system is the delivery method, and the hints are the relationship elements used in the system. The language then dictates how you translate back to your passwords. The system platform has structure that basically does not vary unless you use the position function in some way or you choose to specify the relational hints in another order in your hint script. The system and language will become integrated, but you need to flexibly use what you create. Your system can be positional (first part, second part) or real subject matter relationships in a dynamically scripted order, at the time you build the password or a new password for the same system later. You can randomize the relationships into any positional order you wish and develop a few ways to specify the positions, as we said above. This is not a jumble of letters but a way of ordering elements in a password by some language representation for the password sets. Be flexible! You can also have a theory for jumbling a set of whole passwords.

Here you go:

Debt;Reversed is the hint in a musical theory relationship. The password is "Not;Deeperindebt", from the old song "Sixteen Tons". I know from the ";" that it is music related. Also since the character is placed in the hint as is, it is the separator for the result (a single character function as is), and not specified as the word semi-colon. Why is it so useful to me, and why does it make such a great password relationship. I sang this song as a youngster so many times that it is ingrained in my mind. The words are "you load 16 tons and what do you get, another day older and deeper in debt". The simple theory of reversing the answer is my choice. Do you remember the use of the ";" in chapter 3 for the F Clef? Music theory is the relationship for the entire set of elements in the password.

What could be a variation of this system? How about Reversefirstpart;keepDebt? The answer or password could be "Is;Deeperindebt". This is a word theory relationship in the same system based on the default root represented by Not;Deeperindebt. Not is the first part of the default root. Could you ever figure this out if I did not tell you the relationship? Remember that reversal by itself is a simple theory, and the music theory theme where the parts of the song come into play is the overall concept of the system, which has as many element relationships as I can get out of all the words of the song. Notice that I did not reverse the letters but the meaning. Are you thinking now? Did you notice the capital letters in the hint translating to the capital letters in the password? Keep is a function stating keep the root's second part and make the first letter upper case. Since this is not a positional system literally I use firstpart with the literal ";" separator to tell me it has two parts however. You are going to say what happened to the words of the song. You will find that the literal use of the ";" tells me it would be my music theory system. If it is all based on the song, this works. Another example in the song would be weightverb; with a password result of load;. The words of the song could be used for 12 passwords for the year each month having a default root. Take any of the passwords and establish a root for each and set is as the default then add functions to it to change it. You can use "verb" to audibly communicate the root whatever it is. Since it is a hint, who would know that the password is load;. You can make the root not the smallest character set but load;xx placing a suffix on it. The suffix is your suffix, what ever you normally use.

Let's discuss the ";" a bit more. Note that the key to this whole system is one character, the ";", with so much meaning, yet it is so simple. Use a significant character to cause your mind to trigger a progression of information that is so vast. In my case it works well. You may not know the song and it may also not have as much meaning for you as it does for me. We were a struggling family in my youth, and my father worked hard and I could see the 16 tons on his shoulders, but he never complained.

Now you want to decide on your personal subject matter or related subject matters. Think of something that you know that no one else knew. Best would be a feeling in your mind about an event that was significant in your life. The thought process used in the association should be complex. The password itself does not necessarily have to be character complex.

Consider something about yourself, your brother, or something about a project you once worked on, and no one ever knew that you had a special something happen in that project. My brother passed away, so I can use things that I know were in his mind that I know he did not relate to anyone else. Even if he did, I would make the answers so obscure as to dilute the meaning for others that knew him. He was however very quiet about things. His name was John, and he also, was a hero to me. So not-hero could be answer not-john. Also consider that several hints can mean the same thing. Finally try to have all the hint elements for a password tightly relate.

Don't limit your thinking. As an example, several letters in a word can stand for numbers or visa-versa. I just gave you one way to specify numeric passwords as a word or a progression of letters. You could also have each word mean a relational or translated number. You can convert words to dates as numbers, and then the numbers could have a different set of relationships. It is your choice. This is a double relationship where you would have to translate through two passes of relationship. You would have maybe a single relationship to get the date and then the 8 relationships from the date to get the answer. The date could be a person's birthday and the person's name is the password. This is complex and you could only use short word relationships as space may be a problem in some site-keys. You would have to really need to do this and ultimately it would have to be right on as to the way you thought out the total of 9 relationships.

As long as you write the script it will tell you the answer. If you fail to write the script, you will not remember it unless you memorize the password. And that is not the way to do it.

We now have come to the next rule. Write the script down in you list, along with the username, and never try to memorize a password.

OK, I will give you an alpha to alpha hint system.

I had a computer game I worked on, many years ago. It did something so well in testing it told me that the work product was done so I stopped testing that part of the application.

The event: The artificial intelligence methodology in the programming fooled me into betting half of my money. I had three kings, and the game program had the 4th king (which I did not know). It took no draw of cards. I thought it was bluffing because I was the one that opened (jacks or better) with the three kings. Well, it had a royal straight flush, the highest hand possible. I bet heavily during the entire hand's play as the randomizer faked me out with what I thought was a bluff. Yes, I created my first monster as a scientist at age 17. It just ate me alive, and I was its creator.

So I will choose for the system we create here a subject of royal-flush, which has two parts and a set of hints that would be gamefirst and gamesecond. Wow, positional and relational at the same time. Remember that this is based on an actual event to tie my mind to the relationship that makes me not have to remember. It just releases from my mind and not from my memory. You have to understand and see this intuitively or you will never create a truly fool proof system from your relationships.

Now, lets create a password suffix for them. So I would write a script that says:

Gamefirst lit xxx revmonth julie-julie up. Separate with blanks to make it clearer in your script when you write it down.

So the real password would be Royal100:

The literal (lit) tells us that the three xxx's are literally three digits, not one or two, and the reverse (rev) says not 001 for January, but 100 becoming reversed, so February would be 200. December would be 210. Much of this is your choice. Yes you are creating your own challenge.

If the literal were not in the hint then xxx could mean that the suffix was only one digit, so 0,1,2,3,4,5,6,7,8,9 as opposed to 000,00

1,002,003,004,005,006,007,008,009. Of course the reverse would not be usable in the one digit system because you would have to allow 10-12 taking two digits minimum.

Remember that systems can have exceptions that you specify in some way. Again, your choice (an ex would be the exception function or you can specify it in theory some way). You need to do that with your mind, I can't help you with that.

Complicate it by adding a special character ":", hinted by julie-julie, someone special to me. If the special character were a period "." then only one julie would be in the hint.

You need to know your special character and how your system uses it. Document it then learn the theory of its use. Lastly, you can then literally forget the password. Develop a relationship for as many special characters as you can.

You should write your hints down as not all applications give you a hint space in the site-key. More and more web server applications are doing this now however. Someone must be thinking the same way that I do. Remember, if there is a 1 percent chance of something happening it can happen. I am not here to argue about pure worlds. Hey that could be a relationship and a password (Pure & World)! Sorry, I got emotional about the subject matter! You could maybe use the reading of my book to trigger the relationship. If you now see how the mind plays this out, you are there.

Using some knowledge from prior chapters, we can use this system to really confuse people hacking the code.

Reverse all Gamefirst xx monthly-julie-julie down could be ..01layoR for January.

Game rev second xx julie could be "hsulF01." Flush being the second part of the hand name "Royal Flush".

In the second use of the system we are only reversing the subject, not the entire password. And the suffix is a static 01 because there is no need to have a monthly change. Notice the suffix stayed on the end because it was not included in the reverse. The same went for the special character, as it also stayed on the end. Single Julie is a single period.

You can design the placement of functions like reverse to mean different things.

X's are suffixes and prefixes in one of my systems. A suffix can be numeric or alpha characters such as AAA literal. This AAA could also yield 001 for what ever you want the theory to be (IE: January). Or 100 if reverse were placed after the literal function. The choice is yours. You just need to document and then understand your method. Remember that the relationships are not written down so there is no concern for your documentation meaning anything to anyone. You could choose to make 000 mean 00a, 00b, and so forth for the months of the year or quarters of the year. Successive characters in this mean that it is a suffix just like xxx. Shortly it will be your turn to create "systematic theory". Some of you have created systems but have not created scripts to relate the particular password and have had problems getting the password correct at times. This has probably led you to get the password reset several times.

Suffixes are the hardest to create an association for because they need to change over the range of a year in many cases. A monthly change in a password will require you to create a mathematical progression from 1 to 12 relating to the letters of the alphabet (only one way). There are several ways of doing number to letter and also letter to number translations. You need to create a script for it with an association you can translate. Here is a simple one. Every third letter in the alphabet is a designation of the month of the year, so January is a for 1, February is d for 2, and so on, g is 3, j is 4, m is 5, p is 6, s is 7, v is 8, y is 9 and b is 10, e is 11 and h is 12. the script is stated as "xxxlitalphabet3lettermonth". So knowing you are in March (3) makes you translate to ggg. It has to make sense to you.

You can however relate 12 life experiences and make them your suffixes, requiring a second table of elements. Again, think it through. Other than a need to rotate over a year, don't create many suffix values for single static passwords. It could get you into trouble unless you have a way of stating in code which one you changed it to.

Wow, someone just asked if they could see these suffix translation hints they could figure them out. No matter, they will never get the main body of your hints to the password. The suffixes are only to get by the application requirement for 12 month different passwords. This suffix method is simple. Some applications do not allow this simple change from month to month. If so then use 12 different passwords that are subject matter related! A challenge, but you can do it. Remember "related" and something from your life experience.

Remember three things: that you need to know and understand your subject matter and theory. Make sure your system meets the application's password requirements. Finally, if you have to change something for an application, you will simply create an exception hint modification for the particular password, not change the system. It is all up to you. Drastic changes to another subject matter will be hard to get if you don't write the new script down. So if you do make a complete subject matter change, write down the script for sure.

Remember that your password needs to remain complex to satisfy most applications.

This means the use of letters, numbers, and sometimes capital letters and special characters. If the application changes requirements, you can always adapt later. Remember that you will have the flexibility to do so later if you think it thru now. This means look for flexibility in what you create. Think, think, think.

These systems are primarily for you to figure out and get your passwords correct, not necessarily to keep others from guessing your passwords. Others will have enough trouble as your system will allow you to change any password at any time. You simply need to change the hint string (Site-key text) when you change a password so you don't confuse yourself when you next sign on to a particular application using your system.

Your system needs to be precise and consistent otherwise it could backfire on you. The script language also needs to be consistent.

Remember you must learn and understand your theory and how you position the programming of the hints or "elements".

Ok, you now have some insight into what this is all about. It is a programming language, based on the human ability to create, understand, associate, and never forget how the theory works. If you then develop your system, make extensive use of it, and stretch it to find modifications, variations and exceptions, you will have a system that only you can use to recreate your passwords without even having to remember them.

After a while, if you use a particular password many times, you find that you simply type it naturally. I only use my hints now if I am tired or it is for an application I don't use much. I simply open my code book to look up the application hints in my list. Remember that a password could be composed of more than one element or hint for one or more systems. This is all your design choice. Be flexible and creative but not overly so. You may need 30 passwords but no more that three systems, or two, or maybe one.

Ok, once a password is typed many times for an application, it becomes a secondary association with the screen and the name of the application, down to the colors in the screen. You do this automatically due to repetitive conditioning. It becomes a conditioned response when you see the screen. Have you ever noticed that when an application uses a new main screen or modifies the format of the screen, even just a bit, how you hesitate to put in your password, even to the point of saying, what is my password? It is because the association is broken. You might think it is "being caught off guard". Actually in these cases being caught off guard is not the proper analysis. You are just lost and the relationship just does not come to the surface. Some people can get by this, as they are thinking about the application and not the associated screen at the time of entering the password.

Most of my adult life, I have been programming computers and I have created a rudimentary language within an application more than once for command names. Have fun with your creation. I have also written rudimentary languages for applications I have written.

Notice that I never used the word poker in my hints in the "gamefirst" example. It would be giving away about 90 percent of the relationship. A substitute for "game" as a hint element could better be expressed as the word "play" which would relate to guess what? It would relate to the means by which you make use of the game, pushing the relationship down another level to a point that no one would ever guess you were talking about a card game, as it could be sports (play ball in baseball). Think hard as you have so many rich relationships in your mind by age 17. I am 63 at this writing.

Chapter 6

WRITING DOWN YOUR HINTS

The Language

Ok, now we need to write down our application ID list along with the password hints (elements). It is now the time to understand that if your hint is well formulated you can even write down an abbreviation as the hint. This means you are at the point of creating your script language if you have not already started to tinker with it.

The language can be as simple or as complex as you want or need it to be. Try to be as brief as possible so what you create can fit in most site-key descriptions on the internet, where most applications are loaded these days. You can abbreviate your script hints, and also the spelling of your functions.

My new wife and I speak three letters and immediately know that the password is a complex version of a password system we created which is information about each of us that is alpha and numeric with a suffix and a special character as needed (it has three elements or parts to it, with two subject matter relationships). The abbreviation is for the default for that system, but it does not have to be the default or "root". I also use it as part of a set of passwords that I use with another system concatenated to these passwords for other applications. This protects my other applications from another person accidentally writing down the first system part of a password given to them.

If I use this system for sensitive work that others should not know about I simply use a coded variation so I am in compliance with any license work I do. Remember that personal sign-on(s) do not use work related passwords.

Always remember that these systems are for you, more so to get your passwords correct, but it is also good to have peace of mind that they are diversified, one way of being flexible.

Here are some of the hints for my Passwords:
fmg
fmgxxxxjulienotxx --- remember xxxx in your system could mean two characters 01.
frankvanessaxx
frankvanessa literalxxx
frankvanessaxx secondpartxx – two systems used for one password.

These are all for the same password system with one variation, the last. Can you pick out the exception? Exceptions are usually written with a functional word. Yes "literal" tells us the suffix is a three character numeric, not two or one. The suffix could be alphabetic in your system. You can make the rules in your book! Exceptions are usually fixed value results, but do not have to be.

Wow, how can three characters be the hint for a complex password? Ok, it is because the password is the standard default for the password system. It is the password that all other variations came from. When we speak this hint we know it is the default. In other words, the original (in this case) form of the complex password that was created using the system. You really need to know this password subject matter to be able to use an abbreviation. The main reason for using an abbreviation is so you can speak the password in a crowd to others that need it, usually your life or work partner. In this way you keep the password secret. Again, my wife and I have one we use every day, not concerned that other people are listening.

Sorry, I will not tell you the theory behind this hint system. That would maybe allow you to guess my passwords, although it might take you a lifetime to get them correct! Remember that the knowledge base, for the system, is known only to me and my wife along with the subject matter. The abbreviation is however fmg, which makes the first two script element sets the same password (fmg and fmgxxxxjulienotxx).

You will have to develop a way of writing down your hints so you can use variations of the password without confusing yourself. To do this you need to develop a systematic script language.

An example would be "franknosuffix". This means that the standard password is used without a numeric or alpha suffix. Standards compliance for your system might have a suffix or not. I always use a suffix as the standard. The reason is that the standard root should be complex for numbers and letters. Also, to be able to vary the password from month to month, quarter to quarter or year to year, might become a requirement for the application.

Here is a way of specifying your hints as a simple array language. You can make a list of hints in your book of hints that are not for any particular application. You then write hints in an application as "3,1,5". These hints are in an order on paper, although physically unnumbered on this paper. You simply go to the hint list to see the hints by order. This saves space tremendously in a site-key, but it may force you to pick up your list many times to see the actual hint, and you may have to carry it around or put it in an online spread sheet. What it is good for is providing a way to add another password/ element set to a password that will not fit in the applications site-key using another system, so the hint might be 3,11. Are you starting to think?

Think out of the box as you integrate all the parts of what I suggested you learn (platform, system, subject matter, and language). You now have the power to create something useful for perhaps other applications, other than the creation of password encoding.

Chapter 7

Looking For Cool Subject Matter

Here are some possible subject matters. Integrate them with your experiences.

1. The Internet and the World Wide Web (WWW).
   These are two different things that many people mistake for the same thing. Think about the relationship between the two and do some research on the Internet by using the WWW TCPIP HTTP language. Maybe your hints can be WWW terms for passwords that are Internet based protocol terms. Since the Internet terms are well known, you need to create a second level of complexity by jumbling the passwords in a special way or using a relationship for something you once did or developed on the Internet. The latter is the better method for this system. A secret knowledge base method is far better than risking the use of a simple jumble technique. Jumble is a technique you can use in the system to create variations. It would appear as a function in the language. Please note that variations and exceptions are two very different things. You create many variations as part of a system, and only use exceptions to get around an application's quirks.

2. Movie titles and who you first saw a particular movie with. The movie Clock-Work-Orange, and a wonderful girlfriend,

would be a good example for me. She will go nameless but I will never forget her. Remember to use something about the person, and not the person's name, unless no one else you know knew her. Use the name if it is not so critical an application, but if it is critical use a knowledge relationship (something about the person).

3. Where (the city) a particular famous document was signed or written. Connect the document not to the person that wrote it but what it meant to society as the password. Alternatively consider who wrote the document or better yet the ethnic origin of the writer. This is a very good subject matter that is so vast due to the amount of written world history. Even here tie it to your experience. If you can not, find something better.

4. Famous words spoken by US Presidents. It is the skill with which you choose the relationship that will be critical with this one as many know the verbiage. 35th5thdecadexx is the hint. What is the password for the 5th word spoken by John F Kennedy (35th President) in his statement about space and the end of the 60's decade? The word is "Moon" (man on the moon). I would not tell you the theory of how the hint is created if I were going to use it myself. So now I can not use it and neither can you. You can now consider related subject matter for this system with hints like "courage" with a password element result of "book". In any other context you could never see "book" from the hint "courage". Book written by JFK, "Profiles in Courage".

5. Spoken Languages make a great 2 levels removed subject matter. An example would be the tagalog language mixed with famous sayings. One subject matter element could be the word "sayingmundo", which means world. The password is not "world" but in my relationship it is "oyster", from the English saying "let the world be your oyster". You envision something related to this that was an event, or maybe oysters are just your favorite food. It would have to work for you in some unique way. You should still train yourself to see yourself eating the oysters when you use the script if you are an oyster lover. If not, your rules should apply, and don't use this.

Many of your passwords will secondarily become conditioned responses to the main screen of an application, and you will see it in your mind as you type it. You actually start thinking about it before you type, and continue to see it as you type.

If you don't envision the actual text characters, the association will become weak. What I am saying is that you need to train yourself to envision the concept along with the physical characters a few times. Use this as a backup to solidify the response.

Chapter 8

PUTTING IT ALL TOGETHER

Now you should start to think what subject matter elements you want to put into your system and test your memory of who knows the information and if it is so unique that even the people in you inner circle will not even know it.

Think out of the box and then put it into a box! What? I once developed a set of routines that were what I call closed loop programming analysis routines. The effect is that once you run the input through the routine there is absolutely no chance that you have not found all the properties of the data without even knowing the structure of the data before the routine started. This is something you will need to learn to create a good system. You need to see if the system does all that it is supposed to, without problems.

You may also have to create shared passwords that a few people need to know. Ok, remember that these passwords are for not so critical applications. Never use these particular shared passwords for the most critical applications you have.

I think it is time for you to go off on your own. I suggest you pick up books on encryption and encoding. Learn what it is to cipher, or how number systems like base 16 are represented. This might give you some ideas for creating a numeric password system if you need one. I have a combination lock that has a combination that is a binary progression by coincidence.

Remember that a good application encrypts your password, which will be another level of security. Also remember that this system is for you to decipher your passwords, not to be the world's greatest security expert. I am not and never hope to be.

Ok, remember that you need to open your mind to thinking outside of the box. Be creative, but be consistent with your system and learn how to develop variations and exceptions that are also in keeping with what you normally do within a particular system.

You can combine multiple theories and concepts from the few systems I have discussed into the development of one. It is up to you to create a full system platform. There should be no limit to your thought. You will have to follow the engineering process you are most familiar with to make sure you have a closed loop! That however is another book that would take many more pages.

Be prepared to think about your past, such as school, sports, family, birth country, work experiences, and more.

Most importantly have fun with this, and then enjoy your creation. It will last you a lifetime of passwords that have meaning and are not just randomly chosen. You may only create a few passwords in one system. Most people do not need many.

Thank you papa! I wanted to devote this whole page to my father. The person I loved the most in my life. He was everything to me, my deepest association.

Appendix A

THE SCRIPT LANGUAGE

At this point you have built your system and have chosen your elements or "passwords". Maybe you are here to browse. That is also ok, and a good thing.

If you are here after reading then you should have a good grasp of what is to follow.

Use this Appendix to model your own language to cover your systems.

Remember I said you would not have to understand but a few things. This is an example below. Learn the theory of your own system.

The script format:

I know you secretly always wanted to write your own language (your choice). This is a sample that can have any function substituted for the ones shown. You can create your own functions. I have not included most of mine as they also have association as part of the words, and the spelling would not mean anything to you. Yes, you can integrate subject matter with the function name. This is a sample format and not an actual script.

Prefix-passwordhint1-special-splitliteralxxxx(n1,n2)
suffix-reversepasswordhint2-special-suffixsame

Notice that the example's format differs from the first to the second password set. You can have it the other way around. Passwordhint1 and passwordhint2 are the different subject matters you would have to use in place of the word passwordhint.

The intent here is to show how simple or complex the hint language can be. It is the unusual relationship of the hint to the actual password that is the strongest part of the system. Your language of course has to be clear to you. Remember the old principle of KISS (Keep it simple stupid). Then vary from a good base. You may not be able to get all of this into site-keys so abbreviate in the site key, keep it simple, and refer to your list if you really need to go this far (example above) in creating a script for a password. It is long only to show some of the function possibilities. Normally your hints would be 20 percent of this example's size.

The use of hyphens is optional to separate elements or parts of an element. In this example above, the suffix has two functions which act upon it (split and literal). There is a second relational password as part of the full password, which has a function (reverse). The script could go on for as many passwords in one as you like. You can eliminate the passwords and use prefix and suffix to create a numeric only password with functions acting upon them. Why? Some applications have a, numeric only, password requirement. If you are really cool you can create relationships or positional criteria for prefixes and suffixes for these numeric-only passwords. Think outside of the box and then put it into a box and test it.

The scripts created from this language (in this Appendix) can be both positional and functional. A function appears before the value that it operates upon. This is your choice as with every part of your language.

Hints can be created for all of the elements if you wish.

An element set can have: a prefix, the body of the element that is the strength of the subject matter, a suffix, and the use of special characters that are also hinted at. Remember that special characters and even prefixes and suffixes can also have relationships. It is entirely up to you how you do it. You need to understand that they have to be strong relationships. An example would be the age you considered yourself free, maybe 18 or 21. This might be a good suffix. Just say suffix free in your script.

Within the script are functions (or functional operators that tell you what to do with the password). These functions are something you develop for your system. I will provide a few in this Appendix and in the text of the book.

You can build more functions like jumble for passwords with a numeric number sequence in parentheses following the function "J" or "jum" or "jumble". It is your choice for spelling. If you abbreviate use an upper case letter to tell you it is a function. Put an & before the function as an alternative way of specifying function. It is all your choice.

Other Sample Functions:

Rot or "rotate": An example for rotate would be the relationship to your 6 children each year, every month depending on age as an example. And what you are rotating is not the age or their name but a special relationship you have to each. Wait a minute, there are 12 months in the year. You need two per child. rotateHint-snipxx. You need a list of the 12 relationships written down by month. The "snip" function says no suffix used for this password. You can also use noxx in place of snipxx.

Chain: If written as chain3 this could be the third level of a subject matter. Let us say that (chain3;) is the hint with a ; to signify music. It could mean quarter for quarter note, and it does in this example. The top or first level is a whole note, the next down in the chain is half note and the next, or third down, is quarternote. It is better to use life relationships for things that happened to you or to people you know, than subject matter like music. This is only one example of how to use chain as a theory function. You can also call the function any name you wish. What if you could associate the very subject matter we just used with something in your life? Then it would have associative recall value.

Rev or "reverse": if the hint is revstevenxx (my hero) then the password would be oreh10.

Be flexible with your thinking. Try using functions to specify your suffixes (&SSLxxx).

This could mean shift suffix left for the non-0 digits of letters. So if the suffix you normally use is 001, the result would be converted to 100. ZZA would become AZZ when &SSLxxx is applied. Again, this is all your choice.

Enough example. You should have the idea by now. You can create your own language now. You don't need me any more. My objective has been met. Go be creative.

www.ingramcontent.com/pod-product-compliance
Lightning Source LLC
La Vergne TN
LVHW042138040326
832903LV00011B/292/J